page 1

page 5

page 12

page 8

page 9

page 20

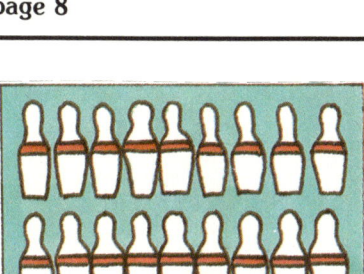

page 23

page 29

page 2

page 21

More stickers at back of book!

Spin and Roll

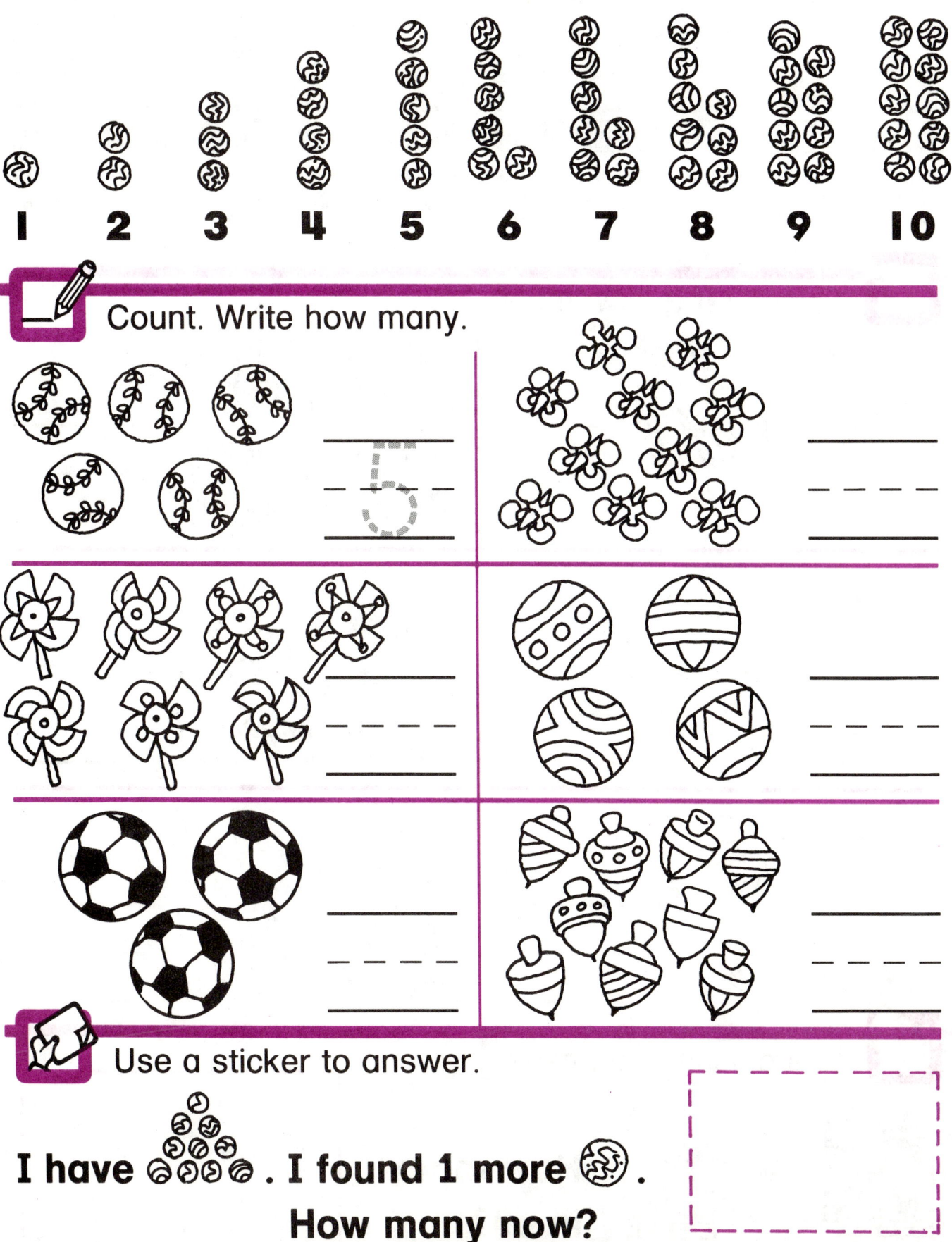

Count. Write how many.

Use a sticker to answer.

I have ... . I found 1 more . How many now?

Skill: To count objects and write numbers to 10.

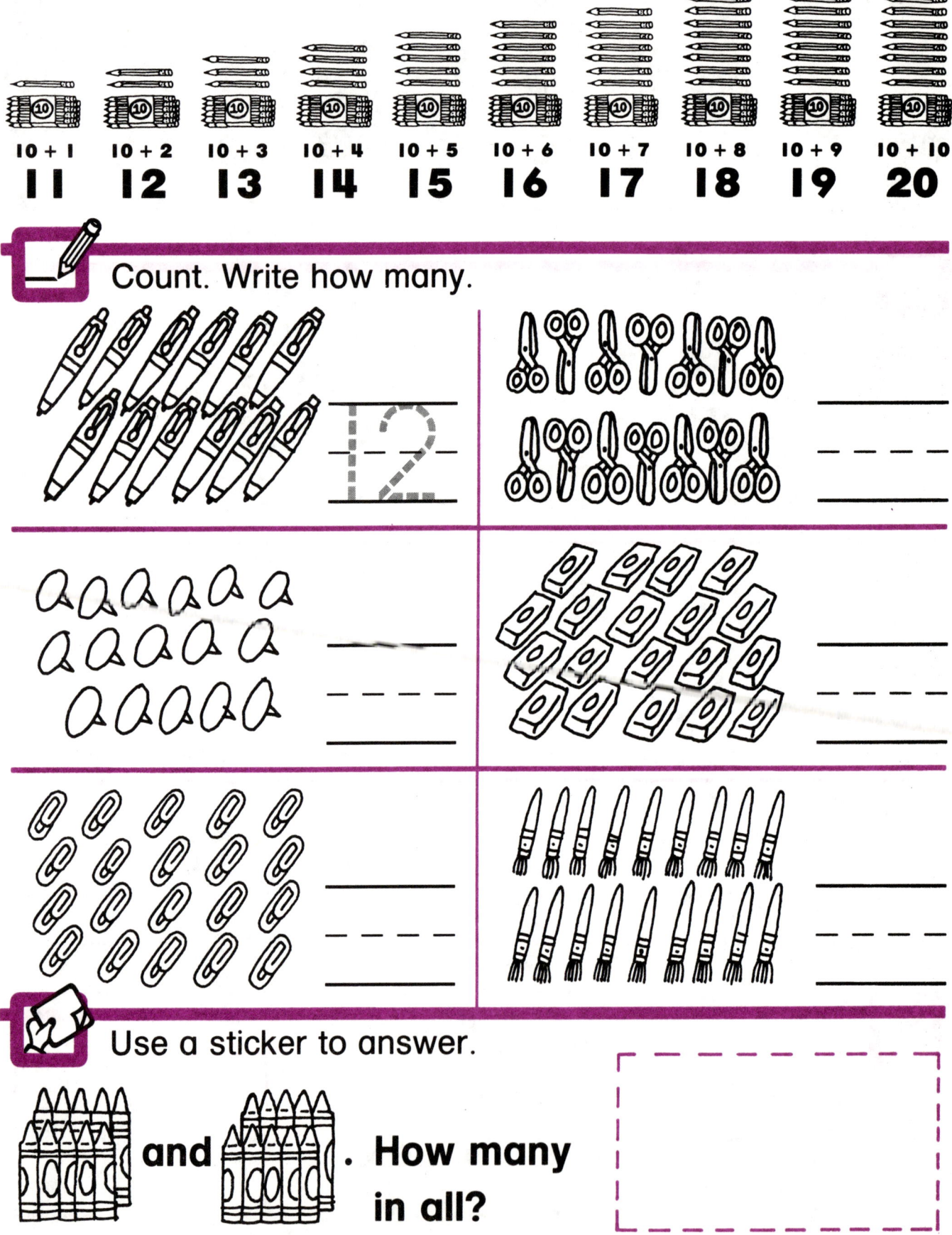

In the Garden

 How many ladybugs? Find the sticker.

```
   6
+  4
────
  10  in all
```

 Write how many.

```
  5
+ 5
───
___  in all
```

```
___
+ ___
─────
___  in all
```

```
___
+ ___
─────
___  in all
```

```
___
+ ___
─────
___  in all
```

Skill: To add numbers to 10.

Shape Match

 Find the sticker that matches each problem.

 Add to find each sum.

$3 + 7 = 10$

$6 + 2 = \underline{}$

$4 + 5 = \underline{}$

$5 + 5 = \underline{}$

```
  3        8        4
+ 4      + 1      + 6
___      ___      ___
```

Skill: To find sums to 10.

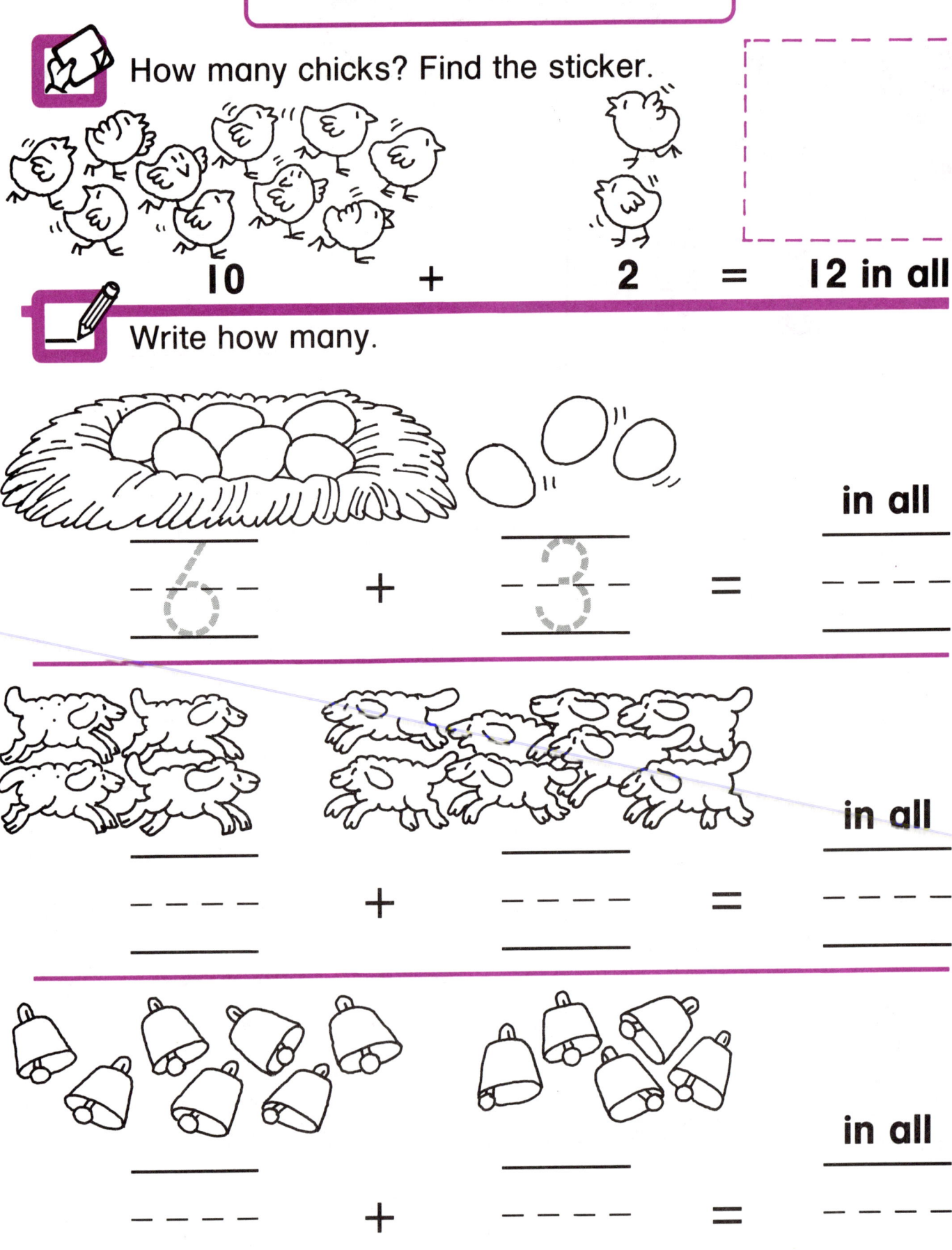

By the Sea

 How many turtles? Find the sticker.

$$\begin{array}{r}3\\+\ 9\\\hline 12\end{array}$$ in all

 Write how many.

$$\begin{array}{r}6\\+\ 5\\\hline \underline{}\end{array}$$ in all

____ in all

$$+$$ ____ in all

____ in all

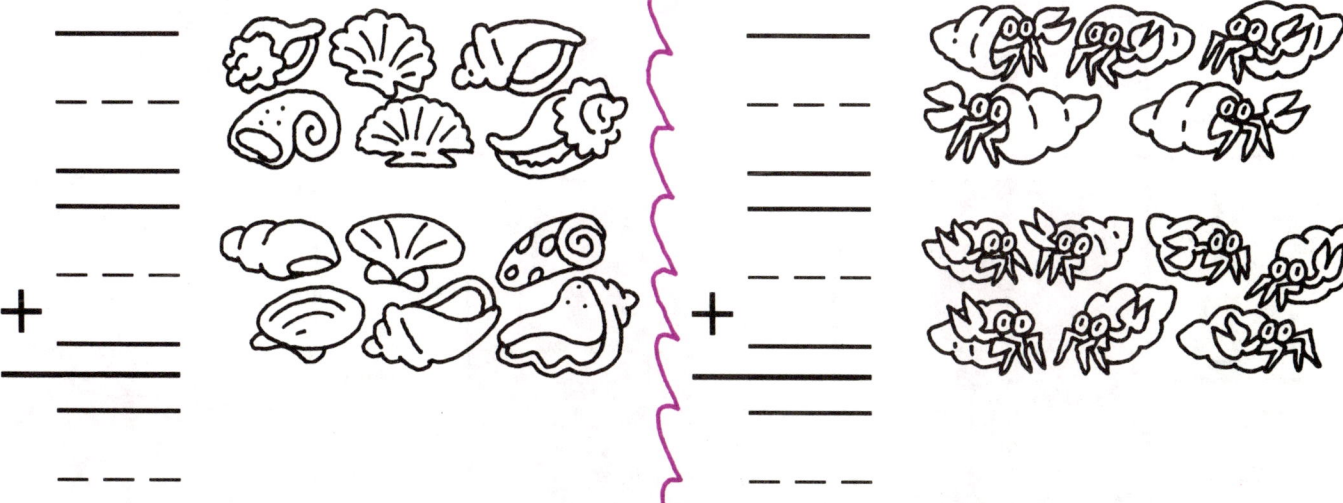

____ in all

Skill: To add numbers to 12.

Domino Dots

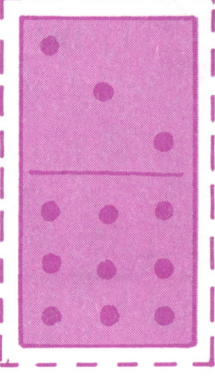

$9 + 3 = 12$

$$\begin{array}{r}3\\+\ 9\\\hline 12\end{array}$$

 Add.

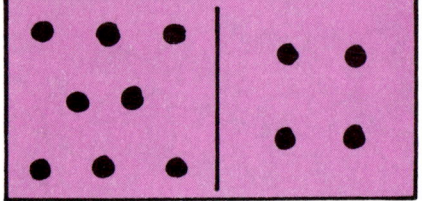

$8 + 4 = 12$ $6 + 5 = \_\_\_$

$$\begin{array}{r}7\\+\ 2\\\hline \end{array}$$
$$\begin{array}{r}5\\+\ 3\\\hline \end{array}$$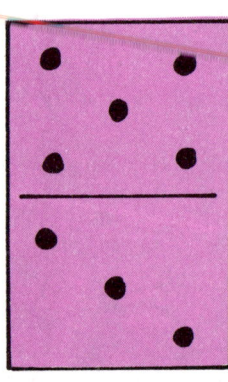
$$\begin{array}{r}2\\+\ 9\\\hline \end{array}$$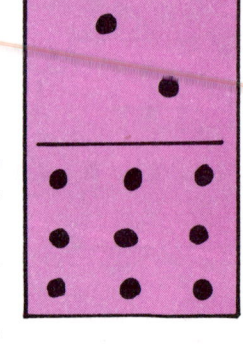

$$\begin{array}{r}1\\+\ 8\\\hline \end{array}$$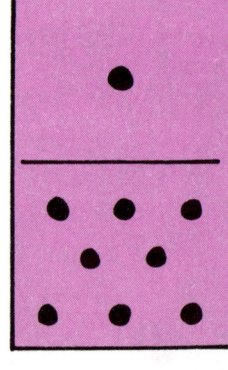
$$\begin{array}{r}4\\+\ 6\\\hline \end{array}$$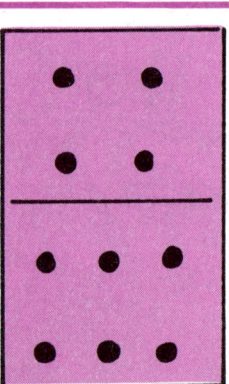
$$\begin{array}{r}5\\+\ 7\\\hline \end{array}$$

Skill: To find sums to 12.

Coloring Fun

 Add. Color the sums.

9 = yellow
10 = red
11 = blue
12 = brown

5 + 7 = ___

4 + 5 = ___

7 + 3 = ___

8 + 3 = ___

$\begin{array}{r} 1 \\ + 8 \\ \hline \end{array}$

$\begin{array}{r} 3 \\ + 9 \\ \hline \end{array}$

$\begin{array}{r} 6 \\ + 4 \\ \hline \end{array}$

$\begin{array}{r} 6 \\ + 5 \\ \hline \end{array}$

2 + 9 = ___

 Use a sticker to answer.

What can go up a  down, but can't go down a  up?

Skill: To solve a puzzle using sums to 12.

In the Pond

How many frogs? Find the sticker.

8 + 7 = 15 in all

Write how many.

6 + 6 = ____ in all

____ + ____ = ____ in all

____ + ____ = ____ in all

Skill: To add numbers to 15.

Flower Show

```
  6              7
+ 9            + 7
----           ----
 15  in all    14  in all
```

✏️ Write how many.

```
   5
+  7
----
----   in all
```

```
+
----   in all        +
                     ----   in all


+                    +
----   in all        ----   in all
```

Skill: To add numbers to 15.

Cars and Blocks

 How many toy cars?

7 + 8 = 15

 Add.

8 + 6 = 14

6 + 7 = ___

9 + 5 = ___   8 + 7 = ___

6 + 8 = ___   7 + 4 = ___

7 + 7 = ___   6 + 9 = ___

```
  9        9        9
+ 3      + 4      + 5
___      ___      ___
```

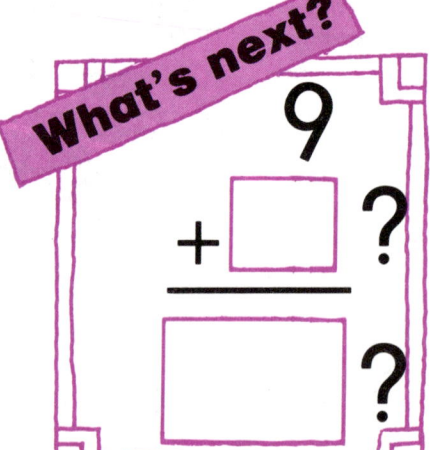

What's next?

9 + ☐ = ?

?

12   Skill: To find sums to 15.

Shoes and Socks

 Add. Find the  stickers with the matching sums. Put them on the socks.

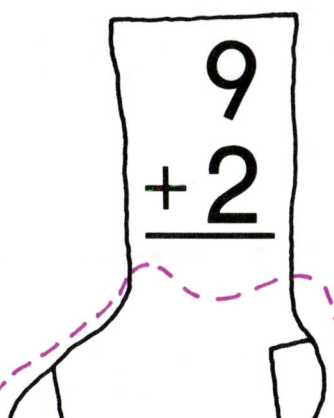

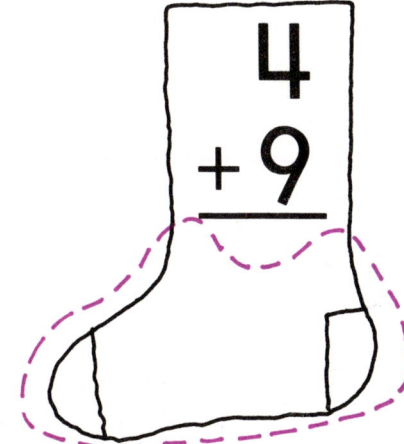

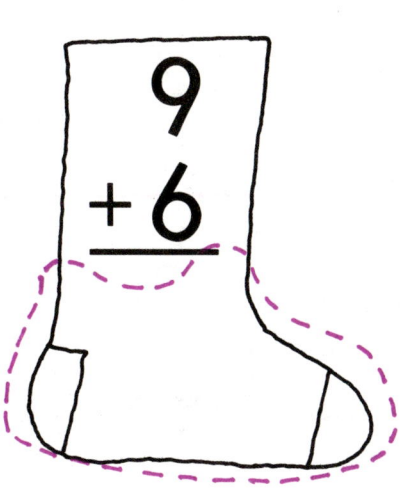

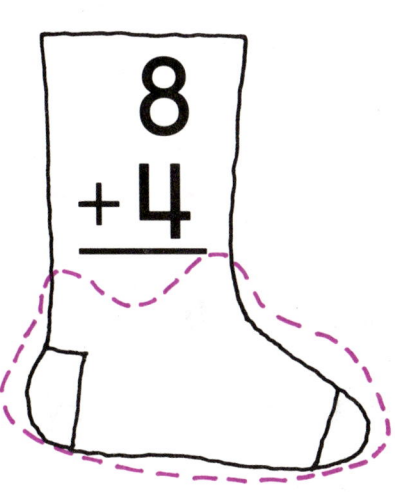

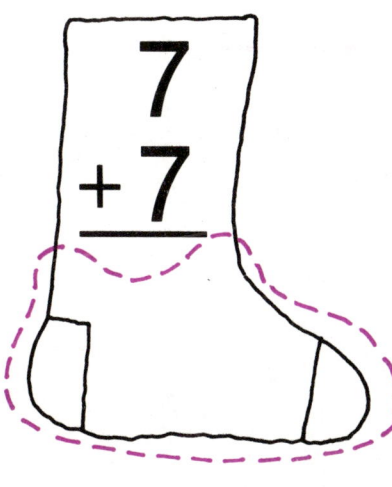

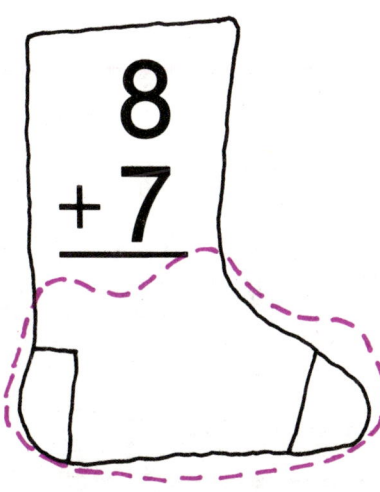

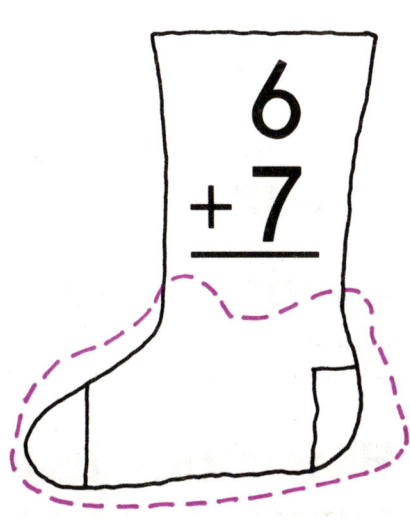

Skill: To find sums to 15.

Flying High

```
  4   kites in the sky
+ 8   more kites fly
 ─────
 12   kites in all
```

✏️ Write the answer.

7 balloons
8 more balloons
_____

How many in all? _____ balloons

---

5 big planes
8 little planes
_____

How many in all? _____ planes

---

9 flags
6 more flags
_____

How many in all? _____ flags

Skill: To solve addition problems.

Stone Soup

Long ago, in a far away land, the people of a small town were very poor.

One day a stranger came to town and asked for food. The people said they had nothing to give him.

But the stranger was very smart.

At last the stranger said, "The soup is done!"

And everyone in the town, including the clever stranger, enjoyed the stone soup.

Recipe for Stone Soup

Pot of water
1 large soup stone
3 large onions
5 fat carrots
10 tiny turnips
8 rabbits
and
lots of imagination

He asked, "Do you have
a big black pot?"
The people said,
"Oh, yes."

He asked, "Do you have
cold, fresh water?"
And the people said,
"Oh, yes."

Just then a hunter came back
from his hunt with 8 rabbits.
They were added to the soup, too.
The stranger stirred the pot for a
very long time.

19 + 8 = 27

The stranger said, "Then fill the pot with water, for I have a soup stone."

"A soup stone? What is that?" asked the people.

"Why, a soup stone makes soup," the stranger said.

So he put 1 stone in the pot of water.

"A turnip or two would really be nice," said the stranger.

"Oh, yes," said the people. And they went and got 10 tiny turnips.

9 + 10 = 19

Then the stranger said, "Do you have an onion or two?"

"Oh, yes," the people said. And they put in 3 large onions.

1 + 3 = 4

"Maybe you have a carrot?" the stranger asked.

"Oh, yes," the people said. And they found 5 fat carrots.

4 + 5 = 9

Sum Money

7¢
+ 8¢

15¢

✎ Write how much money.

9¢
+ 4¢

13¢

8¢
+ 5¢

___¢

6¢
+ 8¢

___¢

7¢
+ 7¢

___¢

✎ Ring all the bags with 15¢.

Skill: To add pennies to sums of 15¢.

Kitten Mischief

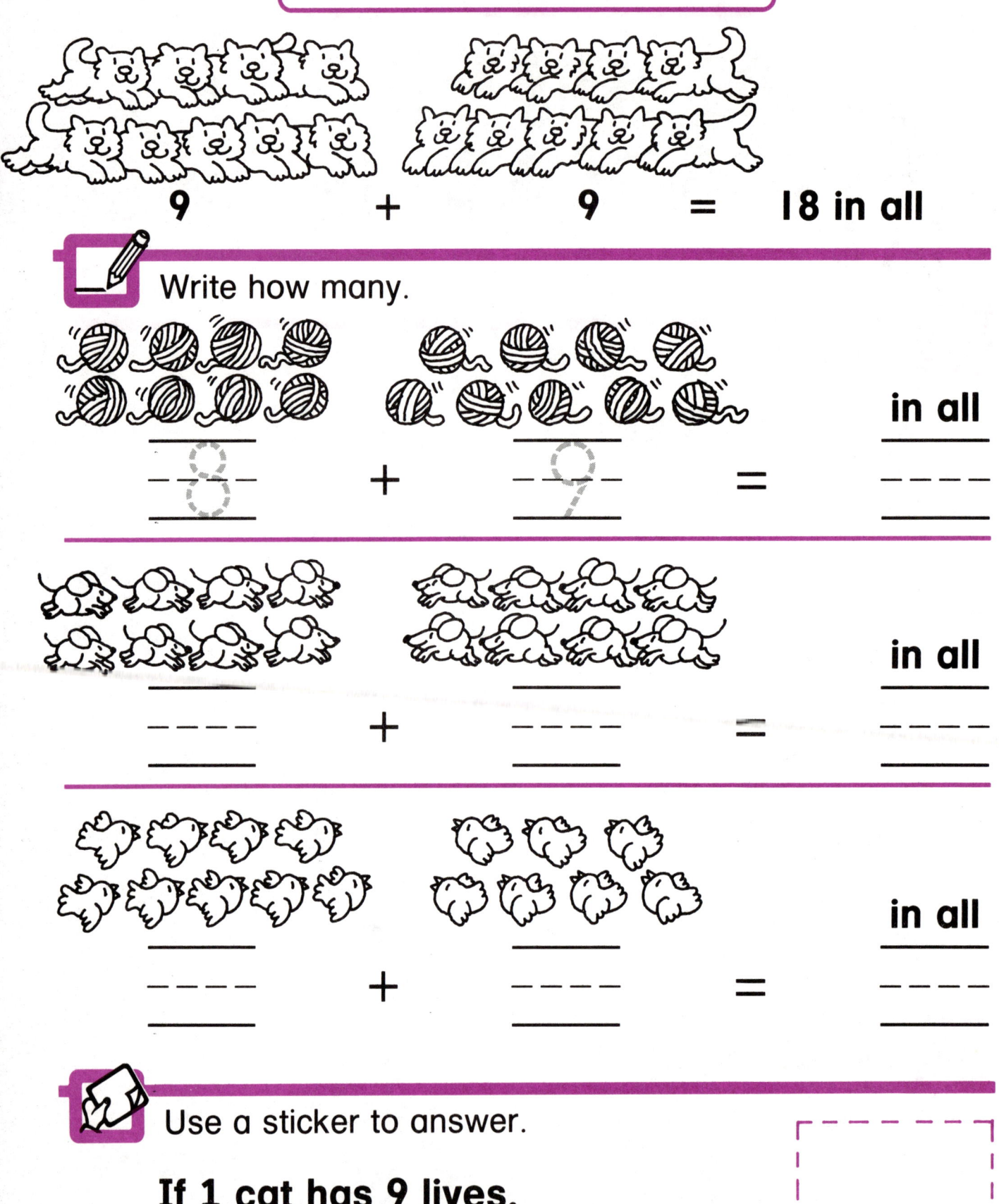

9 + 9 = 18 in all

Write how many.

8 + 9 = ___ in all

___ + ___ = ___ in all

___ + ___ = ___ in all

Use a sticker to answer.

If 1 cat has 9 lives, how many lives do 2 cats have?

Marching Band

```
   9
+  8
-----
  17   in all
```

 Write how many in all.

```
   8
+  7
-----
  15   in all
```

```
   8
+  9
-----
  ___  in all
```

```
   9
+  9
-----
  ___  in all
```

```
   6
+  9
-----
  ___  in all
```

```
   6
+  7
-----
  ___  in all
```

```
   7
+  7
-----
  ___  in all
```

 Use a sticker to answer.

9 . How many ✗?
9 🍗. How many 🍗🍗?

Skill: To add numbers to 18.

On the Green

 How many golf balls?

9 + 9 = 18

 Add.

9 + 8 = 17   9 + 7 = ___

8 + 9 = ___   7 + 9 = ___   9 + 6 = ___

8 + 8 = ___   7 + 3 = ___   0 + 7 = ___

2 + 6 = ___   8 + 0 = ___   6 + 3 = ___

22   Skill: To find sums to 18.

In the Alley

 How many pins?

```
  10
+  8
----
  18
```

 Add.

```
   8        10         9         6
+  9       + 0       + 9       + 8
----      ----      ----      ----
  17
```

```
   6         7         9         8
+  9       + 6       + 5       + 8
----      ----      ----      ----
```

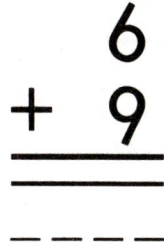

What's next?

```
  10        10        10        10
+  5       + 6       + 7       + ☐ ?
----      ----      ----      
                              ☐ ?
```

Skill: To find sums to 18.   23

Lots of Bugs

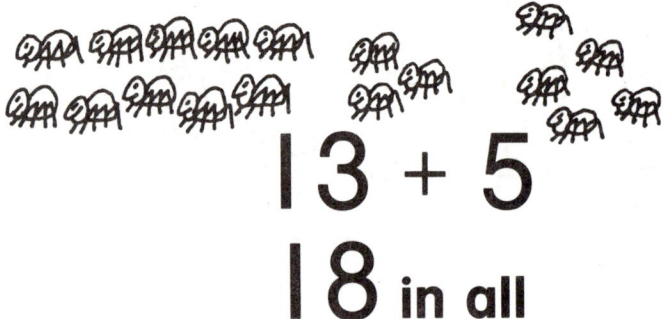

13 + 5
18 in all

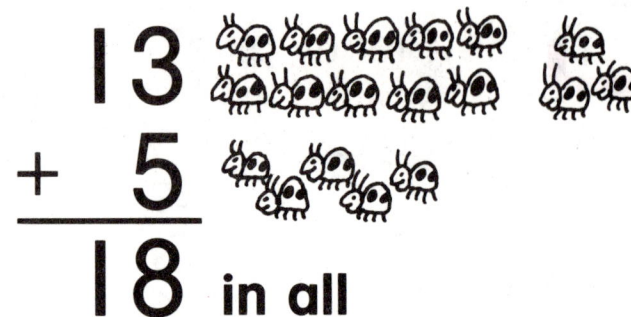

 13
+ 5
---
 18 in all

 Write how many.

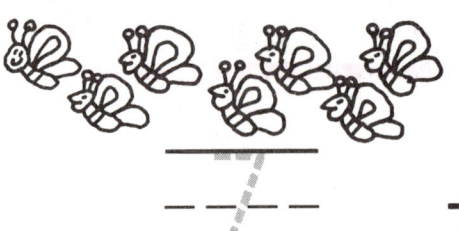  

___7___ + ___12___ = _____ in all

_____ + _____ = _____ in all

_____ + _____ = _____ in all

 Write how many in all.

 14
+ 4

_____ in all

 15
+ 4

_____ in all

Skill: To count to add tens and ones.

Lots of Squares

23
+ 4
---
27 in all

Write how many in all.

13
+ 6
---
19 in all

24
+ 2

___ in all

31
+ 4

___ in all

40
+ 7

___ in all

  5
+22

___ in all

33
+ 3

___ in all

Write how many cents in all.

    = _____ ¢

Skill: To count to add tens and ones.

Fun In Tenstown

In Tenstown everything comes in packs of 10.

tens	ones
3	0
+2	0
5	0

 Add.

tens	ones
4	0
+1	0
5	0

tens	ones
7	0
+2	0

```
  30      10      40      50      60      50
 +30     +70     +30     +20     +30     +40
 ----    ----    ----    ----    ----    ----
```

 Use a sticker to complete.

10 dimes make 1 ☐

If you want one give a holler!

More Fun in Tenstown

Sometimes there are extra ones Tenstown.

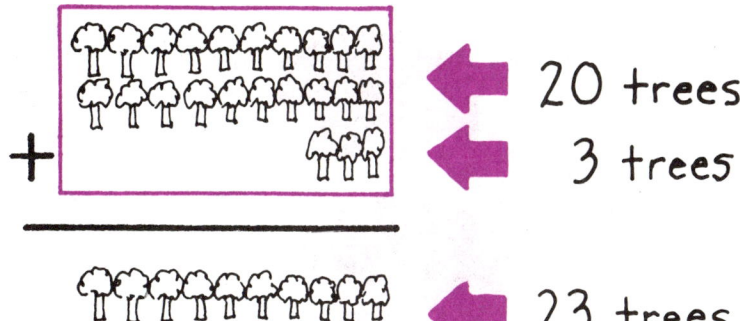

← 20 trees
← 3 trees
← 23 trees

 First add ones. Then add tens.

tens	ones
3	4
+5	2
8	6

tens	ones
2	7
+3	0

```
  51        62        31        17        20
+ 38      + 25      +  7      + 32      + 45
____      ____      ____      ____      ____


   6        71        44        29        36
+ 43      + 26      + 44      + 50      + 12
____      ____      ____      ____      ____
```

Skill: To add 1- and 2-digit numbers without carrying.   27

Add Them All

 Add.

```
  5        4        3        3        3
  3        1        4        2        4
+ 6      + 9      + 8      + 7      + 5
-----    -----    -----    -----    -----
 14       ___      ___      ___      ___


  4        3        2        6        5
  4        6        5        6        2
+ 9      + 7      + 3      + 6      + 6
-----    -----    -----    -----    -----
 ___      ___      ___      ___      ___
```

What's Missing?

7 + 4 = 11

 Write the missing number.

$$\begin{array}{r}6\\+\\\hline 14\end{array}\qquad\begin{array}{r}\\+9\\\hline 12\end{array}\qquad\begin{array}{r}8\\+\\\hline 13\end{array}\qquad\begin{array}{r}\\+9\\\hline 17\end{array}$$

$$\begin{array}{r}\\+8\\\hline 16\end{array}\qquad\begin{array}{r}\\+7\\\hline 7\end{array}\qquad\begin{array}{r}9\\+\\\hline 18\end{array}\qquad\begin{array}{r}10\\+\\\hline 13\end{array}$$

 Use a sticker to answer.

What two numbers can you add and still get 0?

[?] + [?] = [0]

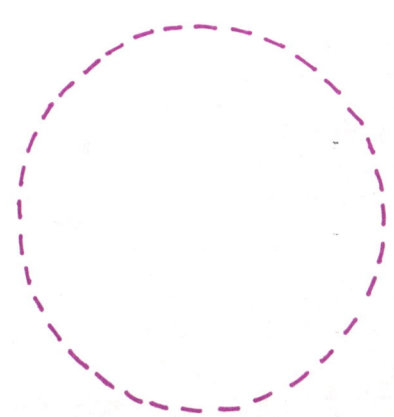

Skill: To find missing addends.

Know It All

 Add.

```
   2        3        5        7        5
 + 7      + 6      + 5      + 6      + 8
 ───      ───      ───      ───      ───
   9
```

```
  50       70       32       65
 +10      +20      + 5      + 3
 ───      ───      ───      ───
```

```
  22       80       35       24
 +30      +14      +22      +61
 ───      ───      ───      ───
```

Practice Test

I can add across.

4 + 3 = ____   ○ 6
● 7
○ 8

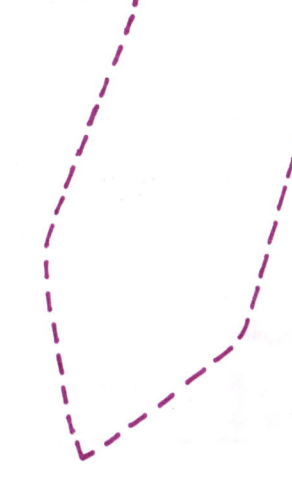

 Add.

A. 3 + 4 = ____
○ 5
○ 6
○ 7

B. 5 + 6 = ____
○ 11
○ 12
○ 13

C. 0 + 6 = ____
○ 0
○ 6
○ 1

D. 7 + 4 = ____
○ 10
○ 11
○ 12

E. 2 + 9 = ____
○ 9
○ 10
○ 11

F. 4 + 9 = ____
○ 12
○ 13
○ 14

G. 8 + 6 = ____
○ 12
○ 13
○ 14

H. 9 + 9 = ____
○ 17
○ 18
○ 19

Skill: To test addition facts to 18.

Practice Test

I can add up and down.

```
  11
+  3
```

○ 12
○ 13
● 14

 Add.

A.
```
  23
+  1
```
○ 23
○ 24
○ 25

E.
```
  40
+ 30
```
○ 7
○ 70
○ 77

B.
```
  65
+  2
```
○ 60
○ 66
○ 67

F.
```
  23
+ 41
```
○ 22
○ 46
○ 64

C.
```
  34
+  4
```
○ 38
○ 48
○ 83

G.
```
   2
   3
+  5
```
○ 10
○ 11
○ 12

D.
```
  60
+ 10
```
○ 7
○ 60
○ 70

H.
```
   4
   4
+  5
```
○ 14
○ 15
○ 13

32 Skill: To test addition skills.

page 4

page 3

page 10

page 7

page 6

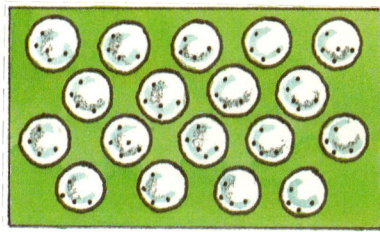
page 22

page 26

Reward Stickers!

page 31   page 32

page 13